Separations Division 2021

Held at the 2021 AIChE Spring Meeting and 17th Global Congress on Process Safety

Online
18 – 23 April 2021

ISBN: 978-1-7138-5337-4

Printed from e-media with permission by:

Curran Associates, Inc.
57 Morehouse Lane
Red Hook, NY 12571

Some format issues inherent in the e-media version may also appear in this print version.

Printed with permission by Curran Associates, Inc. (2022)

For permission requests, please contact AIChE
at the address below.

AIChE
120 Wall Street, FL 23
New York, NY 10005-4020

Phone: (800) 242-4363
Fax: (203) 775-5177

www.aiche.org

Additional copies of this publication are available from:

Curran Associates, Inc.
57 Morehouse Lane
Red Hook, NY 12571 USA
Phone: 845-758-0400
Fax: 845-758-2633
Email: curran@proceedings.com
Web: www.proceedings.com

TABLE OF CONTENTS

619272 Practical Industrial Distillation

Abstract:

This presentation, aimed at young professionals in Separations, provides an overview of industrial distillation - from theory to practice

619277 Introduction to Distillation Column Control

Abstract:

The understanding of the basic control schemes on a distillation column is key for effective operation and troubleshooting. This paper introduces distillation column control, and the concepts behind typical control schemes employed in industrial distillation columns. The advantages, disadvantages, and pitfalls of control configurations will be illustrated using examples.

618908 Guidelines for Column Model Validation and Proper Use of Efficiency in Column Design

Abstract:

Modeling an existing distillation column involves collecting plant data, choosing the best physical property method, setting up the column simulation, and adjusting parameters such as tray efficiency to best represent the column behavior. In this presentation, general guidelines to be followed to successfully model an existing distillation column will be discussed. In addition, definitions and relationship between different types of efficiencies used in trayed column design will be presented. At the end of the presentation, pitfalls of using different types of efficiencies interchangeably in column simulation and design will be illustrated using an example.

619268 Ideal Stage Vs. Rate-Based Simulation of Gas Treating Processes

Abstract:

Rate-based simulation for sour gas treating applications is a relatively new, but proven, technology that was developed in the 1980s and 1990s. The older more traditional technology uses the concept of the ideal and has been used to simulate acid gas removal applications throughout the industry for many years, so what are the advantages to using rate-based simulators? The discussion will cover the conceptual differences and technical bases between mass transfer rate-based simulators and ideal-stage based simulators, why someone would choose one over the other, and the advantages and disadvantages of each.

619274 Introduction to Troubleshooting Distillation Columns

Abstract:

In all chemical plants and refineries there generally is not enough instrumentation in the right places in
our processes. During the design phase, decisions were made between knowing more about the process,
which means buying more instruments, and saving capital funds. When a problem appears with the operation of a vessel, there are always gaps in the knowledge of what is happening due to a lack of instruments in the places where knowledge is not routinely required. Process engineers and operations personnel need to know what is happening so they can take corrective action. When knowledge is missing, they often take actions that make matters worse.

For distillation and separation processes Chemical Engineers have historically relied upon plant process
measurements such as flow rates, temperatures, and pressures and model or simulation results for data with which to solve troubleshooting or re-design problems. An array of on-line diagnostic services are available that can provide additional data offering real-time information on how pieces of process equipment are actually operating. The most common application is Gamma Scanning. This test is primarily applied, but not necessarily limited to, distillation or separation columns. Gamma scans provide a density profile of the internal process of operating distillation columns and other process vessels. The density profile can be used to diagnose the hydraulic operating conditions of mass transfer
devices such as damage to internals, flooding, degree of entrainment or weeping, liquid levels on trays and distributors, liquid distribution through packed beds, etc. The presenter will show case studies where scanning revealed vital process information that helped solve an operating problem, or helped make a revamp/re-design successful. Additional derivatives from gamma scanning of detecting flow distribution patterns will be demonstrated. The ThruVision Scan technique is a specialized horizontal gamma scan used to generate a topographic profile of the internal cross-sectional density of process equipment. This profile is useful for the detailed study of liquid flow distribution through packed columns as well as having applications beyond distillation columns. A hybrid application involving tracers
and scanning on fixed bed catalytic reactors will also be discussed.

NOTES

NOTES

NOTES

NOTES

AUTHOR INDEX